1 Atomic structure

Sub-atomic particles

All atoms are made up of a small, central **nucleus** with tiny sub-atomic (smaller than an atom) particles called **electrons** around it. The electrons are a long way from the nucleus.

The nucleus itself is not a single particle but is made up of two kinds of sub-atomic particles called **protons** and **neutrons**. ▶

The particles have mass and electrical charge. They are measured according to the mass and charge of the proton.

Protons have a mass of 1 and a single positive charge. Neutrons have the same mass as protons but are electrically neutral. Electrons have such a small mass it can be ignored, but have a negative charge equal to the positive charge of the proton.

It is the attraction between positive and negative charges that holds the electrons in place around the nucleus.

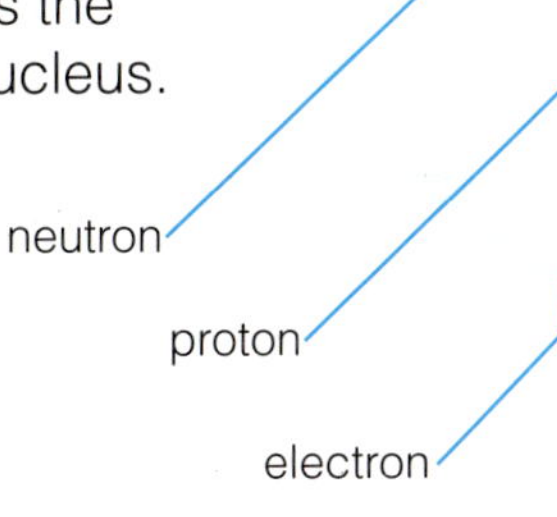

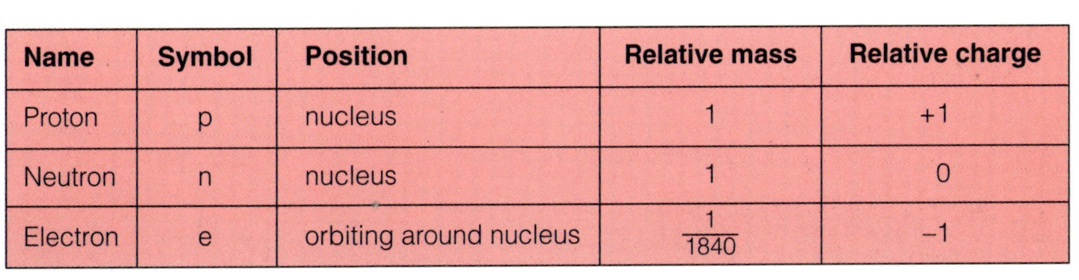

Name	Symbol	Position	Relative mass	Relative charge
Proton	p	nucleus	1	+1
Neutron	n	nucleus	1	0
Electron	e	orbiting around nucleus	$\frac{1}{1840}$	–1

Q1 Name the particles in the nucleus of the atom.

Q2 Name the particle orbiting the nucleus.

Q3 Which particle has a very small relative mass?

Q4 Which particle is neutral?

Q5 Which particle is positively charged?

Q6 Which particle is negatively charged?

Q7 Which two particles are of equal mass?

Q8 Explain why the particles which orbit the nucleus do not fly away from it.

Atomic number and mass number

All atoms are electrically neutral. The number of protons always equals the number of electrons. This number is called the **atomic number (Z)** of the element. It is unique to that element. The number of neutrons in an atom can vary but it will still be neutral. The total number of protons and neutrons in an atom is called its **mass number (A)**.

The rules of atomic structure can therefore be listed as:

- All atoms have a nucleus. This contains protons and neutrons.
- Electrons orbit around the nucleus.
- The number of protons = the number of electrons.
- The number of protons = the atomic number (Z).
- The number of protons + the number of neutrons = the mass number (A).

Let us look at atoms of aluminium:

- aluminium atoms have 13 protons and 14 neutrons in the nucleus
- because there are 13 protons in the nucleus there must be 13 orbiting electrons
- because there are 13 protons in the nucleus the atomic number (Z) of aluminium is 13
- all atoms of aluminium must have 13 protons and no other element has an atomic number of 13
- because there are 13 protons and 14 neutrons in the nucleus the mass number (A) of the atom is 27

the atom can be written as

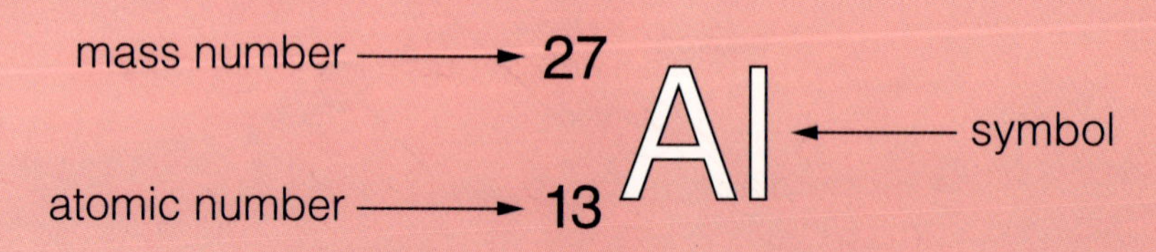

Q1 What is meant by atomic number?

Q2 What is meant by mass number?

Q3 Explain why all atoms are electrically neutral.

Q4 An atom has 23 protons in its nucleus.
a How many electrons does the atom have?
b What is the atomic number of the atom?
c If the mass number of the atom is 48 how many neutrons does the atom hold?

Q5 An atom has a mass number of 64 and an atomic number of 29.
a How many protons does the atom have?
b How many neutrons does the atom have?
c How many electrons does the atom have?

Q6 How many protons, neutrons and electrons does each of these atoms have?

a $^{16}_{8}O$ **b** $^{39}_{19}K$ **c** $^{235}_{92}U$ **d** $^{1}_{1}H$

Extension exercise 1 can be used now.

The arrangement of the electrons

The electrons in an atom orbit the nucleus in a series of **shells** or **energy levels** which are at different distances from the nucleus.

Each shell has a maximum number of electrons that it can hold. ▶

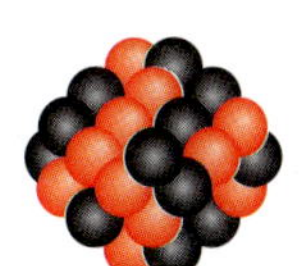

Let us look again at the aluminium atom

$$^{27}_{13}Al$$

- ☐ because the atom has 13 protons in the nucleus it must have 13 orbiting electrons
- ☐ these electrons will be placed 2 in the first shell, 8 in the second shell and 3 left over in the third shell
- ☐ the **electronic configuration** of Al is thus **2.8.3**
- ☐ the complete atomic structure of $^{27}_{13}Al$ can now be seen as

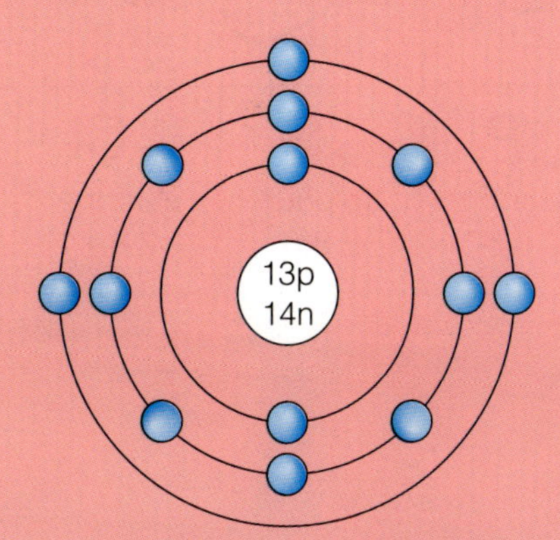

The shells are filled with electrons in order, starting with the first shell. Each shell must be filled before the next one is used. ◀

Q1 What is meant by electronic configuration?

Q2 A sodium atom can be written as $^{23}_{11}Na$.
a What is the atomic number of sodium?
b How many electrons are there in the sodium atom?
c What is the electronic configuration of sodium?
d What is the mass number of sodium?
e How many neutrons are there in the sodium atom?

Q3 Draw a sketch showing the complete atomic structure of the sodium atom.

Extension exercise 2 can be used now.

2 The Periodic Table

The first section

When the electronic configurations of the elements are compared to their chemical properties, we see that elements which have the same number of electrons in the outer shell have similar properties.

For example lithium (2.1), sodium (2.8.1), and potassium (2.8.8.1) are metals with very similar patterns of behaviour. These metals are called the **alkali metals**.

Two further examples are the **halogens** – fluorine, chlorine, bromine and iodine – which each contain seven electrons in the outermost shell; and the **noble gases** – helium, neon, argon, krypton and xenon – which each contain eight electrons in the outermost shell.

The Periodic Table lists the elements in order of their atomic number. Elements having the same number of electrons in their outermost shell are put in vertical columns.

The Russian chemist Dmitri Mendeléev first proposed in 1869 that the elements could be arranged in a pattern which he called the Periodic Table. ▲

The first 20 elements look like this ▼

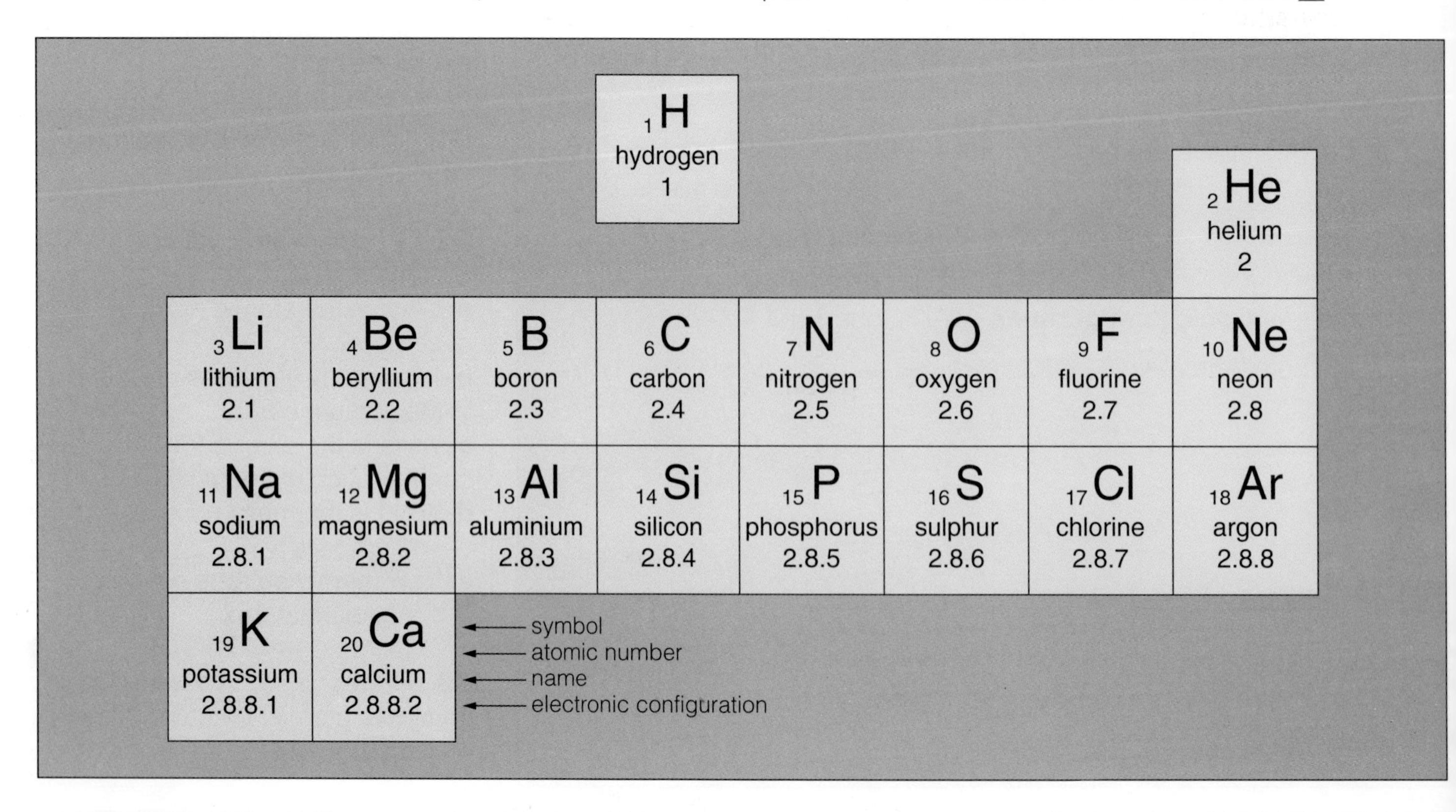

			$_{1}$H hydrogen 1				
							$_{2}$He helium 2
$_{3}$Li lithium 2.1	$_{4}$Be beryllium 2.2	$_{5}$B boron 2.3	$_{6}$C carbon 2.4	$_{7}$N nitrogen 2.5	$_{8}$O oxygen 2.6	$_{9}$F fluorine 2.7	$_{10}$Ne neon 2.8
$_{11}$Na sodium 2.8.1	$_{12}$Mg magnesium 2.8.2	$_{13}$Al aluminium 2.8.3	$_{14}$Si silicon 2.8.4	$_{15}$P phosphorus 2.8.5	$_{16}$S sulphur 2.8.6	$_{17}$Cl chlorine 2.8.7	$_{18}$Ar argon 2.8.8
$_{19}$K potassium 2.8.8.1	$_{20}$Ca calcium 2.8.8.2						

Using this arrangement of elements which we call the Periodic Table it is possible to make predictions about elements with which we might not be familiar. For instance, the full list of alkali metals is:

All of these metals have atoms with one electron in the outer shell. They are therefore very similar to each other. Knowledge about how lithium, sodium and potassium behave enables us to make predictions about the behaviour of francium. In Chapter 3, which looks at the alkali metals, you will be asked to make such predictions.

Lithium	Li
Sodium	Na
Potassium	K
Rubidium	Rb
Caesium	Cs
Francium	Fr

Similarly the full list of halogens, all of which have 7 electrons in the outer shell, includes the element astatine, At. In Chapter 4, which looks at the halogens, you will see that we can make predictions about astatine by studying the behaviour of the other halogens fluorine, chlorine and iodine.

After the first twenty elements the Periodic Table becomes more complicated. The electrons begin to use the sub-shell of the third shell. This fills from 8 to 18 electrons over the next 10 elements. These elements are the **transition elements**. They are all metals and they fit in the table in transition between the second column (Group II) and the third column (Group III) of the Periodic Table.

Q1 Copy and complete each of the sentences below.

a In the Periodic Table the elements are arranged in order of their ____________________ .

b The halogens have ____________________ electrons in their ____________________ .

Q2 Name three of the alkali metals. Give the electronic configuration of each. Explain why they have similar properties to each other.

Q3 Explain why beryllium, magnesium and calcium are together in the same column of the Periodic Table.

Q4 Which elements are called the noble gases. What do their electronic configurations have in common?

The full Periodic Table

- ☐ The table has eight groups of elements plus a block of transition metals.
- ☐ The zig-zag line through the table separates the metals and the non-metals. The metals are left of the line.
- ☐ The vertical columns are called **groups**. A group contains elements which have the same number of electrons in their outermost shell and therefore have similar properties to each other.
- ☐ With the exception of the noble gases, which have eight electrons but are in Group 0, the group number is the number of outermost electrons that the elements in that group have.
- ☐ The horizontal rows are called **periods**. Period 1 contains only hydrogen and helium. Period 2 contains the elements lithium to neon. Period 3 contains sodium to argon, and so on.
- ☐ The atoms of the transition metals have more complex electronic configurations. Some of the common metals such as iron, copper and zinc, are transition metals (see Extension exercise 4).

1 H hydrogen

period \ group	I	II			the transition metals		
2	3 Li lithium	4 Be beryllium					
3	11 Na sodium	12 Mg magnesium					
4	19 K potassium	20 Ca calcium	21 Sc scandium	22 Ti titanium	23 V vanadium	24 Cr chromium	25 Mn manganese
5	37 Rb rubidium	38 Sr strontium	39 Y yttrium	40 Zr zirconium	41 Nb niobium	42 Mb molybdenum	43 Tc technetium
6	55 Cs caesium	56 Ba barium	57 La lanthanum (58–71)	72 Hf hafnium	73 Ta tantalum	74 W tungsten	75 Re rhenium
7	87 Fr francium	88 Ra radium	89 Ac actinium (90–103)				

58–71 – elements 58 (cerium) to 71 (lutetium)

(90–103) – elements 90 (thorium) to 103 (lawrencium)

- ☐ From left to right across a period we see a gradual change in properties from the most metallic elements in Group I to the most non-metallic elements in Group VII.
- ☐ Moving down a group of metals there is a gradual increase of metallic properties. Moving up a group of non-metals we see a gradual increase of non-metallic properties.
- ☐ Thus it can be seen that the most metallic elements are at the bottom left-hand corner of the table, while the most non-metallic elements are at the top right-hand corner.
- ☐ In this book and its extension exercises we shall be investigating some of these concepts more closely by looking at the chemistry of the Group I elements (the alkali metals), the Group VII elements (the halogens), the Group 0 elements (the noble gases) and some of the transition metals.
- ☐ We shall also look at some of the chemistry of aluminium (Group III), carbon (Group IV) and nitrogen (Group V).

					III	IV	V	VI	VII	0
										2 He helium
					5 B boron	6 C carbon	7 N nitrogen	8 O oxygen	9 F fluorine	10 Ne neon
the transition metals					13 Al aluminium	14 Si silicon	15 P phosphorus	16 S sulphur	17 Cl chlorine	18 Ar argon
26 Fe iron	27 Co cobalt	28 Ni nickel	29 Cu copper	30 Zn zinc	31 Ga gallium	32 Ge germanium	33 As arsenic	34 Se selenium	35 Br bromine	36 Kr krypton
44 Ru ruthenium	45 Rh rhodium	46 Pd palladium	47 Ag silver	48 Cd cadmium	49 In indium	50 Sn tin	51 Sb antimony	52 Te tellurium	53 I iodine	54 Xe xenon
76 Os osmium	77 Ir iridium	78 Pt platinum	79 Au gold	80 Hg mercury	81 Tl thallium	82 Pb lead	83 Bi bismuth	84 Po polonium	85 At astatine	86 Rn radon

each set is a transition series fitting within a transition series

KEY

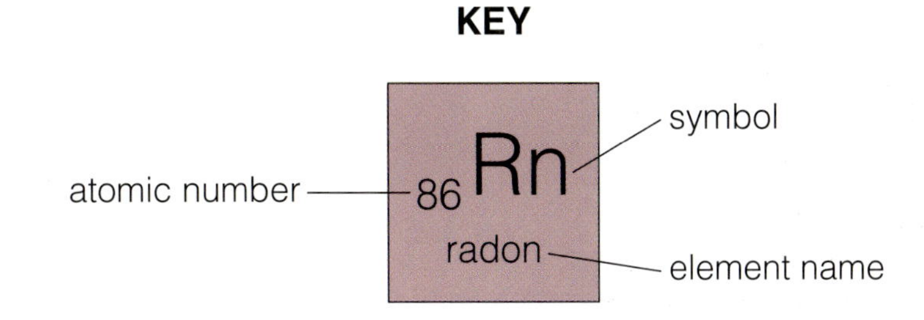

Q1 Give the name and symbol of each of the Group VII elements.

Q2 What is the collective name for the Group VII elements?

Q3 Give the name and symbol of four transition metals from Period 4.

Q4 Give the name and symbol of two elements you might expect to have similar properties to each of:
a calcium
b oxygen
c aluminium.

Q5 Where in the Periodic Table would you expect to find:
a the most non-metallic of the elements
b the most metallic of the elements
c elements with atomic number from 3 to 10
d elements with four electrons in the outermost shell?

Extension exercises 3 and 4 can be used now.

3 Group I – the alkali metals

Physical properties

The alkali metals are difficult and dangerous to handle.
Your teacher may show you some samples.

Name	Symbol	Atomic Number	Electronic Configuration	Density (g/cm³)		Melting Point (°C)		Boiling Point (°C)	
Lithium	Li	3	2.1	0.53		180		1336	
Sodium	Na	11	2.8.1	0.97		98		883	
Potassium	K	19	2.8.8.1	0.86	Increase	63.5	Decrease	757	Decrease
Rubidium	Rb	37	2.8.18.8.1	1.53		39		697	
Caesium	Cs	55	2.8.18.18.8.1	1.88		29		670	
Francium	Fr	87	2.8.18.32.18.8.1	?		?		?	

Lithium. ▲

Sodium. ▲

Potassium. ▲

The alkali metals:

- ☐ are soft enough to be easily cut with a knife
- ☐ are shiny when freshly cut
- ☐ have low densities, increasing down the group
- ☐ have low melting points, decreasing down the group
- ☐ have low boiling points, decreasing down the group.

It should be noted that as we move down the group the atomic number increases and the atom gets bigger, so:

- ☐ the density increases because there is more mass in each atom
- ☐ the melting point and boiling point decrease because the bigger atoms are easier to separate from each other by heating.

Q1 Because francium is a highly radioactive element which decays very quickly, little is known about it.
Predict:
a its density
b its melting point
c its boiling point.

Q2 Explain why the melting points of the alkali metals decrease as the atomic number increases.

Q3 Explain why the densities of the alkali metals increase as the atomic number increases.

Reactions of the alkali metals with air

1 If left **exposed** to the air, lithium, sodium and potassium become corroded. Rubidium and caesium catch fire.

◀ Sodium corrodes when exposed to the air for a short period of time.

The alkali metals are very reactive. The metals react with the oxygen, water vapour and carbon dioxide in the air. For example,

sodium + oxygen + water + carbon dioxide → hydrated sodium carbonate

$$4Na(s) + O_2(g) + 2H_2O(g) + 2CO_2(g) \rightarrow 2Na_2CO_3.H_2O(s)$$

This becomes coated on to the piece of sodium.

2 All the alkali metals will **burn** in air, reacting with the oxygen to form a white powdery oxide. For example,

sodium + oxygen → sodium oxide

$$4Na(s) + O_2(g) \rightarrow 2Na_2O(s)$$

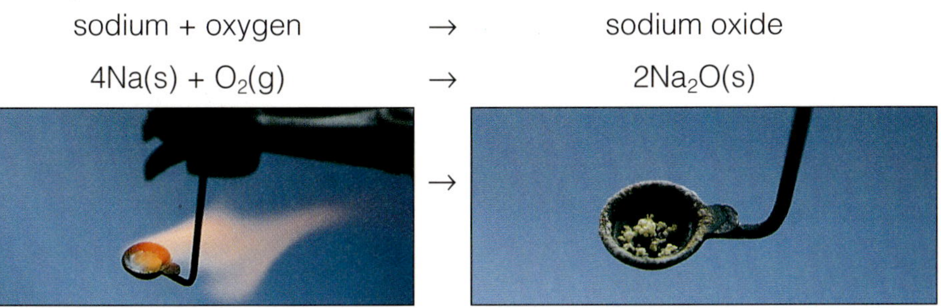

Sodium burns in air with a yellow flame. ▲

White sodium oxide powder is produced. ▲

Lithium begins to burn only after it melts. Sodium and potassium burn while still solid. Rubidium and caesium burn spontaneously on exposure to air.

These reactions of the alkali metals with air show that reactivity increases as we move down the group and as the atomic number and size of atom increase. ▼

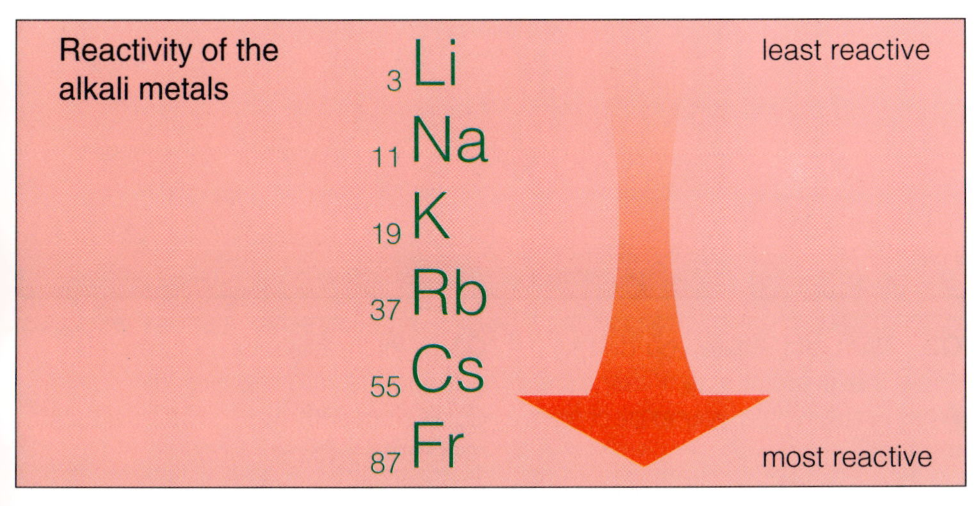

Q1 What is the name and colour of the product formed when potassium burns in air?

Q2 Write a word equation for the burning of potassium in air.

Q3 Write a balanced chemical equation for the burning of potassium in air.

Q4 Rubidium and caesium both react spontaneously when exposed to the air. Which one would you expect to react the faster? Explain your answer.

Reactions of the alkali metals with water

All the alkali metals react easily with cold water. The products are hydrogen gas, which is given off into the air, and an alkaline solution of the hydroxide.

For example,

lithium + water → lithium hydroxide + hydrogen

$$2Li(s) + 2H_2O(l) \rightarrow 2LiOH(aq) + H_2(g)$$

These reactions can be violent and need to be done very carefully. Your teacher may demonstrate them.

Lithium floats and reacts quietly in cold water. ▲

Sodium melts to shiny globules which move about the surface of the water. ▲

Potassium melts to shiny globules but the heat of the reaction is enough to set fire to the hydrogen produced. A lilac flame is seen. ▶

Rubidium and caesium react even more violently than potassium. These reactions are too dangerous to show.

Again we see that the reactivity of the alkali metals increases with increasing atomic number.

Q1 When alkali metals react with water the solution produced is 'alkaline'. What does this mean?

Q2 What test would you do to show that a solution is alkaline? What result would you expect from the test?

Q3 Write a balanced chemical equation for the reaction of potassium with water. Name all reactants and products.

Reactions of the alkali metals with halogens

The alkali metals react with the Group VII halogens to form a metal halide.

For example,

sodium + chlorine → sodium chloride

$2Na(s) + Cl_2(g) \rightarrow 2NaCl(s)$

These reactions are very vigorous.

Sodium burns with a bright orange flame in chlorine. ▲

White sodium chloride (common salt) crystals are formed. ▲

- Lithium reacts more slowly with chlorine than sodium does but potassium reacts much more quickly.
- With bromine, lithium reacts slowly, sodium reacts moderately, and potassium reacts explosively. For example,

potassium + bromine → potassium bromide

$2K(s) + Br_2(l) \rightarrow 2KBr(s)$

Again we see the increasing reactivity of the alkali metals as atomic number and size of atom are increased.

Q1 Write a balanced chemical equation for the reaction of lithium with fluorine. Name the reactants and products.

Q2 Explain why the chemical reactions of the alkali metals are the same throughout the group.

Q3 Write the formulas of the oxide, hydroxide and chloride of rubidium.

Extension exercises 5 and 6 can be used now.

Reactions of the alkali metal compounds

The compounds of the alkali metals are stable and not dangerous. We can investigate some of their properties in the laboratory. We can use sodium compounds which are typical of all alkali metal compounds.

Q1 Copy this table.

Name of material	Formula of material	Appearance	Does it dissolve in water?	Approx pH of solution	Effect of adding silver nitrate solution	Effect of adding barium chloride solution
sodium chloride						
sodium carbonate						
sodium sulphate						
sodium hydroxide						

Apparatus

- ☐ 100 cm³ beaker ☐ spatula
- ☐ test tube rack ☐ sticky labels
- ☐ three test tubes
- ☐ universal indicator paper and pH chart
- ☐ dropper bottles of dilute hydrochloric acid (irritant), nitric acid, silver nitrate solution (corrosive), barium chloride solution (harmful)
- ☐ pure samples of powdered sodium chloride, sodium carbonate, sodium sulphate
- ☐ about 50 cm³ of dilute sodium hydroxide solution (irritant)
- ☐ eye protection

Wear eye protection when handling chemicals.

A Look at the sample of sodium chloride. Note its appearance.

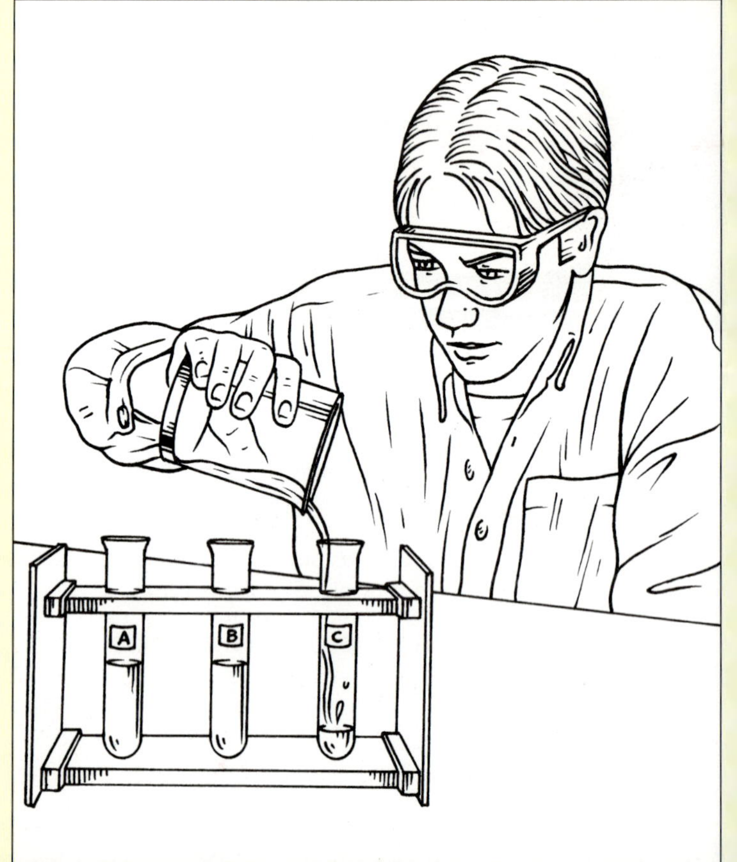

B Put about four spatulas of sodium chloride into a 100 cm³ beaker. Add about 50 cm³ of water and stir the mixture well. Look to see if the sodium chloride dissolves. ▲

C Put three test tubes in a test tube rack. Label them A, B, and C. Divide the contents of the beaker roughly equally among the test tubes. ▲

D Test the contents of test tube A with universal indicator paper. ▼

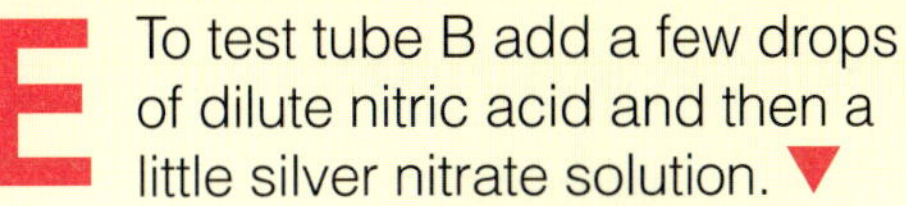

E To test tube B add a few drops of dilute nitric acid and then a little silver nitrate solution. ▼

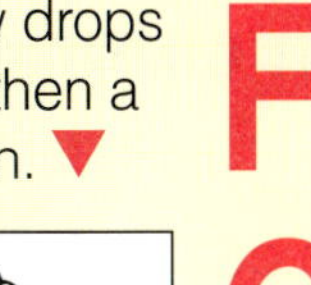

F To test tube C add a few drops of hydrochloric acid and then a little barium chloride solution.

G Complete row 1 of your table.

H Repeat **A** to **F** with:

a sodium carbonate

b sodium sulphate.

Complete rows 2 and 3 of your table.

I Your teacher will show you some solid sodium hydroxide pellets. Use a ready-made solution of sodium hydroxide for **C** to **F**. Complete row 4 of your table.

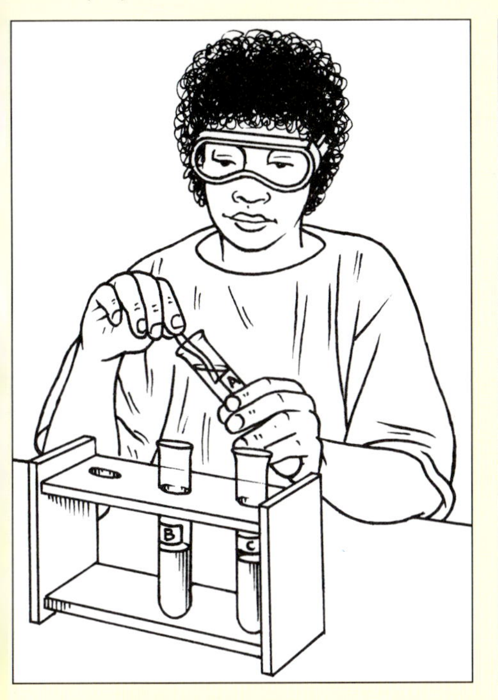

Q2 Identify each of the sodium compounds A, B, C, D from the information given about it.

A This sodium compound is strongly alkaline in solution.

B This sodium compound produces a white solid when acid and silver nitrate solution are added to its solution.

C This sodium compound produces a white solid when acid and barium chloride solution are added to its solution.

D This sodium compound is weakly alkaline in solution.

Q3 Which sodium compounds were soluble in water?

Q4 Which sodium compound is used to make:

a glass
b soap
c paper
d hydrogen
e baking powder
f sodium hydroxide
g synthetic fabrics
h detergents
i ceramics
j food preservatives?

Uses of sodium compounds

Sodium chloride (common salt)

- ☐ is used to produce sodium hydroxide, hydrogen and chlorine
- ☐ is used in the manufacture of sodium carbonate (washing soda) and sodium hydrogen carbonate (baking powder and indigestion tablets)
- ☐ is used in food as a preservative and to give our bodies essential sodium
- ☐ is used to melt ice on the roads.

Sodium carbonate (washing soda)

- ☐ is used in the manufacture of glass
- ☐ is used to soften hard water.

Sodium hydroxide (caustic soda)

- ☐ is used in the manufacture of soaps and detergents
- ☐ is used in the manufacture of synthetic fabrics
- ☐ is used in paper making
- ☐ is used in the making of ceramics
- ☐ is used in the dyeing industry to clean the cloth.

Extension exercise 7 can be used now.

Electrolysis of sodium chloride solution

Some liquids will conduct electricity. When electricity flows through a liquid causing a chemical reaction to occur, the process is called **electrolysis**.

The electrolysis of sodium chloride solution produces sodium hydroxide solution, hydrogen, and chlorine. Each of these is an important material which is used in large quantities to manufacture other materials.

The chlorine gas produced is toxic.

Your teacher may demonstrate the experiment. ▼

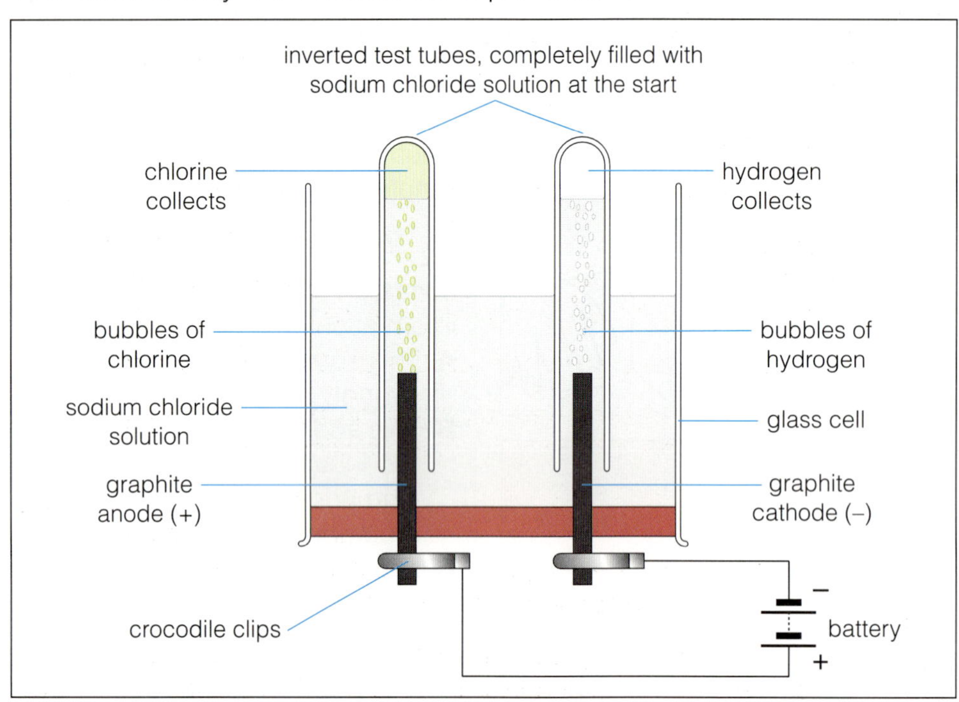

The chemistry of the electrolysis of sodium chloride solution:

sodium chloride solution → sodium hydroxide + hydrogen + chlorine

$$2NaCl(aq) + 2H_2O(l) \rightarrow 2NaOH(aq) + H_2(g) + Cl_2(g)$$

- ☐ The sodium hydroxide solution remains in the beaker.
- ☐ The hydrogen gas is produced at the cathode (–).
- ☐ The chlorine gas is produced at the anode (+).

Chemical tests for each product

The sodium hydroxide is tested with universal indicator paper and is strongly alkaline. ▲

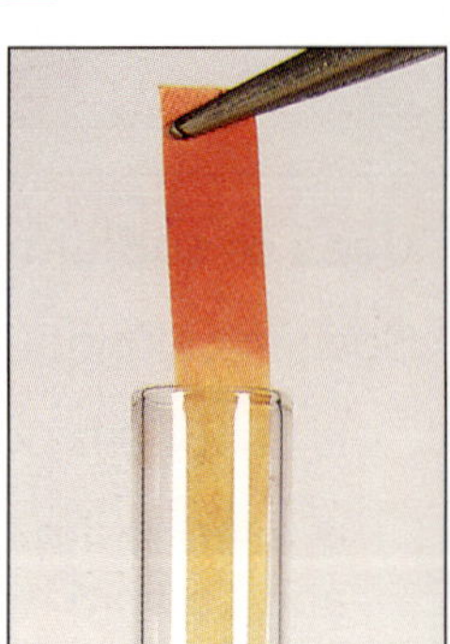

The test tube of chlorine gas is tested with damp litmus paper. It is **bleached** by the chlorine. ▶

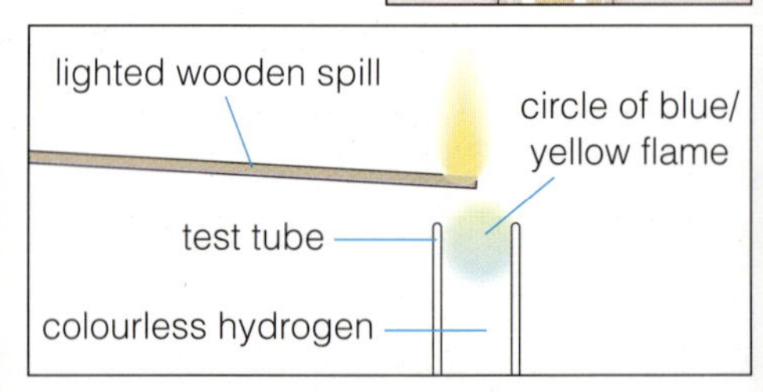

The test tube of hydrogen is tested with a flame. It burns with a 'pop' and a blue/yellow flame. ▲

Q1 What is electrolysis?

Q2 Write a balanced chemical equation for the electrolysis of sodium chloride solution.

Q3 Name the gas produced at the anode.

Q4 Name the gas produced at the cathode.

Q5 Name the material remaining.

Q6 What is the test for chlorine gas? What is the result of the test?

Q7 What is the test for hydrogen gas? What is the result of the test?

Q8 Name the products when you electrolyse a solution of:
a lithium chloride
b potassium bromide.

Extension exercise 8 can be used now.

4 Group VII – the halogens

Physical properties

Name	Symbol	Atomic Number	Electronic Configuration	Colour and physical state at room temp (darker ↓)	Melting Point (°C) (increase ↓)	Boiling Point (°C) (increase ↓)
Fluorine	F	9	2.7	pale yellow gas	−220	−188
Chlorine	Cl	17	2.8.7	greenish-yellow gas	−101	−35
Bromine	Br	35	2.8.18.7	dark red liquid	−7	59
Iodine	I	53	2.8.18.18.7	shiny purple solid crystal	114	187
Astatine	At	85	2.8.18.32.18.7	?	?	?

◀ Fluorine.

Chlorine. ▶

◀ Bromine.

Iodine. ▶

The halogens:

- ☐ are all coloured. The colour gets darker as you go down the group.
- ☐ have very low melting points and boiling points. Both of these increase as you go down the group because atomic number increases and the atoms become heavier.
- ☐ are not all in the same physical state at room temperature:
  - fluorine and chlorine have boiling points below room temperature and are yellowish, toxic gases with choking, irritating smells;
  - bromine has a melting point below room temperature and a boiling point above room temperature. It is a dark red liquid with an unpleasant and toxic vapour. It is one of only two elements that are liquid at room temperature (mercury is the other);
  - iodine has a melting point and boiling point above room temperature. It is a purple solid which produces a vapour which is toxic in high concentration;
  - astatine is a radioactive element of short half-life and little is known about it.

Your teacher may show you some liquid bromine in a stoppered bottle and some iodine crystals.

Q1 What happens to the colour of the halogens as you go down the group?

Q2 Explain why the melting points and boiling points of the halogens increase as you go down the group.

Q3 Very little is known about the radioactive element astatine. Predict:
a its colour
b its melting point
c its boiling point.

Q4 What do you think will be the physical state of astatine at room temperature? Explain your answer.

Extension exercise 9 can be used now.

Chemical reactions and uses of the halogens

1 Reaction with alkali metals of Group I

Each of the Group VII halogens reacts with each of the Group I alkali metals to form metal halides.
For example,

potassium + bromine → potassium bromide

$$2K(s) + Br_2(l) \rightarrow 2KBr(s)$$

The reactions are very vigorous; for example, the potassium/bromine reaction is explosive.

2 Reaction with hydrogen

Each halogen will combine with hydrogen to form a hydrogen halide. For example,

hydrogen + fluorine → hydrogen fluoride

$$H_2(g) + F_2(g) \rightarrow 2HF(g)$$

This reaction is explosive, even in the dark.

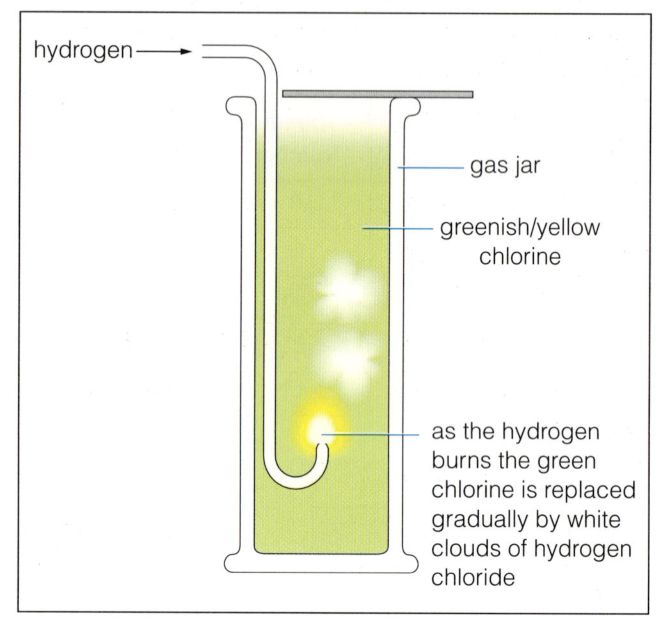

Chlorine will explode violently with hydrogen when exposed to sunlight but a jet of hydrogen will burn safely in an atmosphere of chlorine. ▲

$$H_2(g) + Cl_2(g) \rightarrow 2HCl(g)$$

Hydrogen and bromine vapour burn with a weak flame when lit but will combine smoothly when passed together over a catalyst of platinised mineral wool heated to about 250°C: ▶

$$H_2(g) + Br_2(g) \rightarrow 2HBr(g)$$

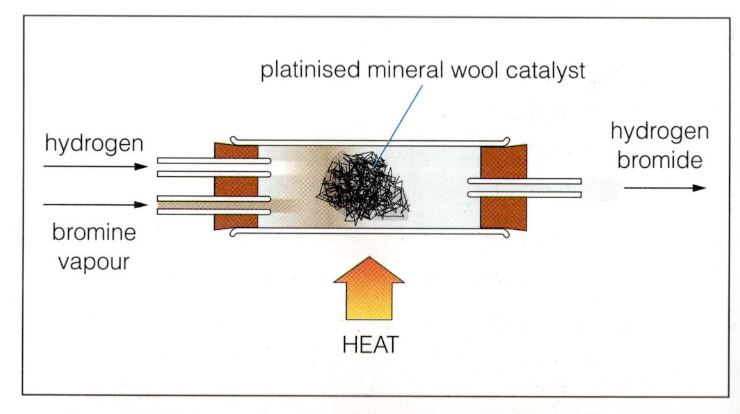

Hydrogen and iodine react in the same way as hydrogen and bromine but the platinised mineral wool needs to be heated to 400°C and, even then, the reaction is not complete:

$$H_2(g) + I_2(g) \rightarrow 2HI(g)$$

3 Comparing the reactivity of the halogens

We have seen that the halogens become less reactive as we move down the group, i.e. as atomic number increases. ▶

Note that this is the opposite of the trend in Group I where the alkali metals become more reactive with increasing atomic number.

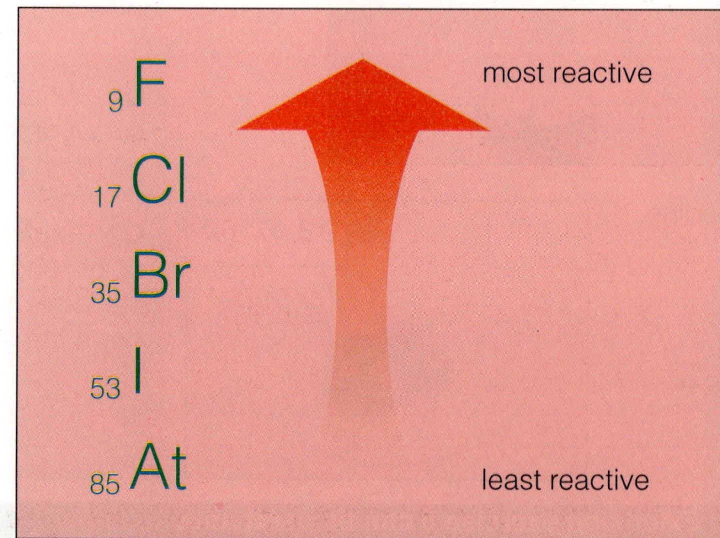

4 Displacement

The halogens become less reactive as we move down the group. This can be demonstrated by a series of displacement reactions.

Chlorine passed into potassium bromide solution releases reddish-brown liquid bromine. ▶

$$Cl_2(g) + 2KBr(aq) \rightarrow KCl(aq) + Br_2(l)$$

Chlorine passed into potassium iodide solution releases dark brown solid iodine.

$$Cl_2(g) + 2KI(aq) \rightarrow 2KCl(aq) + I_2(s)$$

Bromine solution poured into potassium iodide solution releases purple solid iodine. ▶

$$Br_2(l) + 2KI(aq) \rightarrow 2KBr(aq) + I_2(s)$$

These reactions are similar to the reactivity series of metals in which the more reactive metal displaces the less reactive.

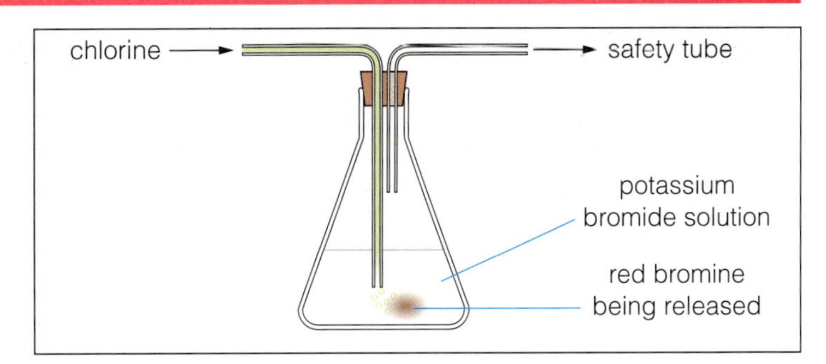

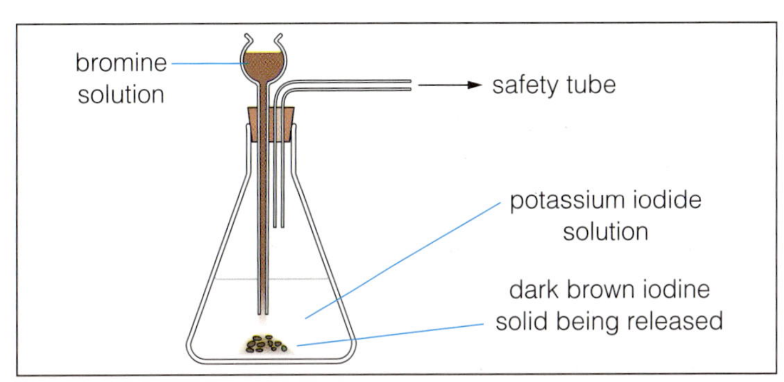

Uses of the halogens

- A lot of chlorine is used to manufacture hydrochloric acid. The chlorine is burnt in hydrogen to form hydrogen chloride

  $$H_2(g) + Cl_2(g) \rightarrow 2HCl(g)$$

  and the hydrogen chloride is then dissolved in water to form hydrochloric acid, a common and cheap industrial acid.
- Chlorine kills bacteria and is used to purify drinking water and water in swimming pools.
- Chlorine is used in the manufacture of bleaches.

  The chlorine is bubbled into water and forms a mixture of hydrochloric acid and hypochlorous acid. This solution is very acidic and a strong bleach. It will quickly and easily remove the colour from many fabrics and papers.
- Fluorides such as sodium fluoride NaF, are added to drinking water and to toothpaste.

  The fluoride combines with calcium phosphate in tooth enamel and strengthens the enamel.
- Iodine is an antiseptic.

  Both antiseptics and disinfectants kill germs but antiseptics can be applied to the skin. A solution of iodine in water can be painted on to cuts and open wounds to kill bacteria.
- Silver bromide and silver iodide are used in photography.

  The silver halide is mixed with gelatine and this forms the film. On exposure to light, metallic silver is produced which, when developed, forms an image on the film.

Q1 Write balanced chemical equations to show how chlorine reacts with:
a lithium
b hydrogen
c sodium iodide solution.
Name all the reactants and products.

Q2 Describe laboratory experiments which show that:
a chlorine is more reactive than bromine
b bromine is more reactive than iodine.
In each case describe what is seen and give a balanced chemical equation.

Q3 Outline, with chemical equations where appropriate, how chlorine is made into hydrochloric acid.

Q4 Give one similarity and one difference between antiseptics and disinfectants.

Extension exercises 10 and 11 can be used now.

Acids

All the hydrogen halides dissolve in water to form solutions which are acidic. The most common of these is hydrochloric acid. This is formed by dissolving hydrogen chloride gas in water. Acids are very common chemical compounds and there are many, many acids. Some are strong acids and need to be handled with great care. Others are weak, harmless acids and can be found in food.

Properties of acids

All acids have these properties:

1 Acids affect indicators

Acids are soluble in water and the solutions formed give characteristic colours with indicators. Litmus goes red in acids and universal indicator gives an orange/red colour depending on the strength of the acid and its pH.

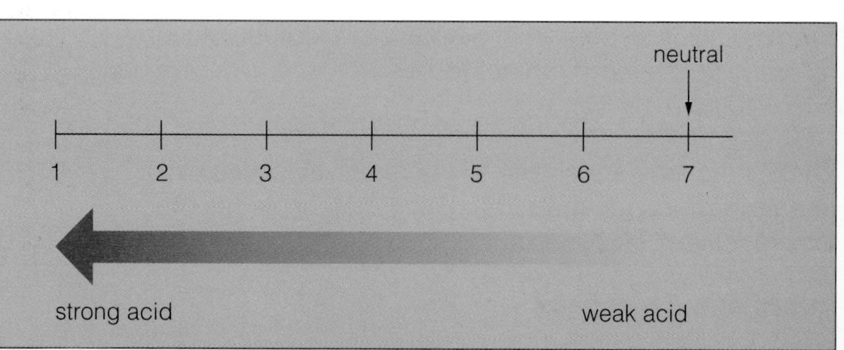

2 Acids and metals

When dilute acids are added to metals, a salt is formed and hydrogen is given off.

A salt is a compound formed when the hydrogen of an acid is replaced by a metal.

Example:

zinc + hydrochloric acid → zinc chloride + hydrogen

$Zn(s) + 2HCl(aq) \rightarrow ZnCl_2(aq) + H_2(g)$

This is a general reaction for all acids and all metals above copper in the reactivity series.

ACID + METAL → SALT + HYDROGEN

3 Acids and bases

When dilute acids are added to bases, which are oxides and hydroxides of metals, a salt is formed and the other product is water.

Example:

magnesium oxide + sulphuric acid → magnesium sulphate + water

$MgO(s) + H_2SO_4(aq) \rightarrow MgSO_4(aq) + H_2O(l)$

This is a general reaction for all acids and all bases.

ACID + BASE → SALT + WATER

4 Acids and carbonates

When dilute acids are added to carbonates and hydrogen carbonates a salt is formed and the other products are water and carbon dioxide.

Example:

sodium carbonate + hydrochloric acid → sodium chloride + water + carbon dioxide

$Na_2CO_3(s) + 2HCl(aq) \rightarrow 2NaCl(aq) + H_2O(l) + CO_2(g)$

The general reaction is:

ACID + CARBONATE → SALT + WATER + CARBON DIOXIDE

Investigating the properties of acids

Q1 Copy this table.

Experiments conducted with dilute hydrochloric acid		
Name of material	**Observations during experiment**	**Chemical equation and names of materials involved**
Universal indicator		
Magnesium		
Copper oxide		
Calcium carbonate (marble chips)		

Apparatus

- ☐ test tube rack ☐ spatula ☐ four test tubes
- ☐ dilute hydrochloric acid (irritant) ☐ dropper
- ☐ wooden spill ☐ copper oxide powder
- ☐ Bunsen burner ☐ test tube holder
- ☐ universal indicator paper and pH chart
- ☐ dilute sulphuric acid (corrosive)
- ☐ strip of magnesium (flammable)
- ☐ rubber bung to fit test tubes ☐ marble chips
- ☐ lime-water ☐ eye protection

Eye protection must be worn at all times.

A Test your dilute hydrochloric acid with universal indicator paper. Note the colour and pH in line 1 of your table.

B ▲ Put about 3 cm depth of dilute hydrochloric acid into a test tube. Add a piece of magnesium ribbon and put the bung loosely into the top of the test tube for about 10 seconds. Remove the bung and hold a lighted spill at the mouth of the test tube. Complete the table.

C ▲ Put one spatula of copper oxide powder into a test tube. Carefully add 3 cm depth of dilute hydrochloric acid. Using a test tube holder, heat the test tube gently over a medium flame until the copper oxide has dissolved. Keep the test tube moving gently during heating. Note your observations.

D ◀ Put one or two marble chips into a test tube. Carefully add 3 cm depth of dilute hydrochloric acid. Hold a pipette full of clear lime-water in the test tube but NOT in the acid. Do NOT squeeze the lime-water out of the pipette. Note your observations.

E Complete column 2 of your table.

Q2 Nitric acid is a very corrosive and dangerous acid. What do you think its pH will be? Explain your answer.

Q3 Soda water is carbonic acid. What do you think its pH will be? Explain your answer.

Q4 Write balanced chemical equations and name the products for each of the following reactions:

a magnesium and dilute sulphuric acid

b zinc oxide and dilute hydrochloric acid

c potassium carbonate and dilute sulphuric acid.

Q5 The salts formed from hydrochloric acid (HCl) are chlorides. Which salts are formed from:

a sulphuric acid (H_2SO_4)

b nitric acid (HNO_3)

c carbonic acid (H_2CO_3)

d hydrobromic acid (HBr)?

Q6 Write a definition of an acid.

F Copy out another table with the heading 'Experiments conducted with dilute sulphuric acid' and repeat **A** to **D** with dilute sulphuric acid. Complete all of your second table.

Extension exercise 12 can be used now.

5 Group 0 – the noble gases

Physical properties

Name	Symbol	Atomic Number	Electronic Configuration	Density (g/dm³)		Melting Point (°C)		Boiling Point (°C)	
Helium	He	2	2	0.18	↓	−271.4	(at 30 atm) ↓	−269	↓
Neon	Ne	10	2.8	0.90		−248.7		−246	
Argon	Ar	18	2.8.8	1.78	Increase	−189.2	Increase	−186	Increase
Krypton	Kr	36	2.8.18.8	3.74		−157		−153	
Xenon	Xe	54	2.8.18.18.8	5.90		−112		−108	
Radon	Rn	86	2.8.18.32.18.8	9.96	↓	−71	↓	−62	↓

The noble gases:

- have very low densities which increase down the group as atomic number increases and the atom gets heavier
- have very low melting points and boiling points which increase down the group as atomic number increases and the atom gets heavier. All the boiling points are well below 0°C and so all of them are gases at room temperature
- all have electronic configurations which have 8 electrons (2 in the case of helium) in the outermost shell
  - all therefore have electronic structures which are stable
  - all are monatomic, i.e. they exist as single atoms and not as diatomic molecules like the majority of other gaseous elements.

Extraction from air

All the noble gases except radon are present in very small amounts in the atmosphere. The concentrations in the atmosphere in parts per million are:

helium	5.2
neon	18.0
argon	9300.0
krypton	1.0
xenon	0.08

Radon can be found in rocks. It comes from the radioactive decay of uranium. Radon is radioactive and has a very short half-life.

Helium is used in weather balloons and airships. ▲

Neon is used in advertising signs. ▲

Argon is used in household light bulbs. ▲

Krypton and xenon are used in high powered lamps. ▲

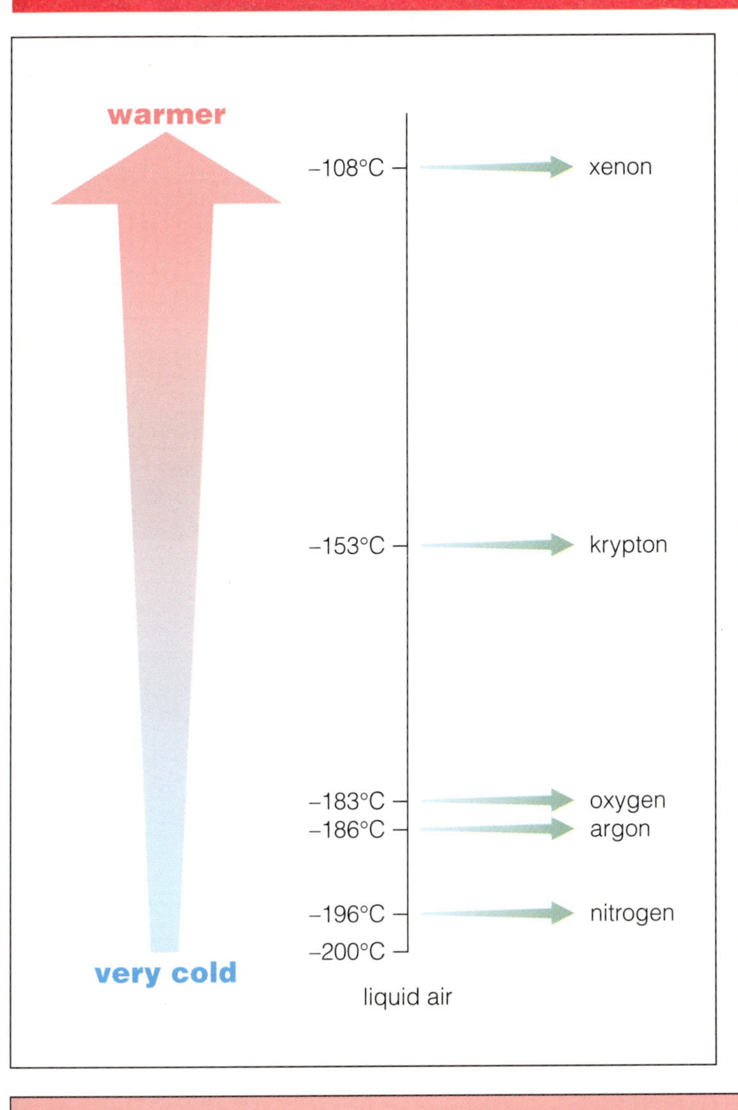

All the noble gases in the atmosphere can be obtained from the **fractional distillation** of **liquefied** air.

The air is filtered and compressed in stages, cooling between each stage. Water and carbon dioxide are removed and at −200°C all the air is liquefied except helium and neon which can now be removed.

◀ As the air is allowed to warm up from −200°C the different gases boil off separately as their boiling point is reached. Oxygen and nitrogen are also obtained during this process.

Chemical reactivity of the noble gases

The noble gases each have a complete outermost shell of electrons. For helium this is two; for the other noble gases it is eight.

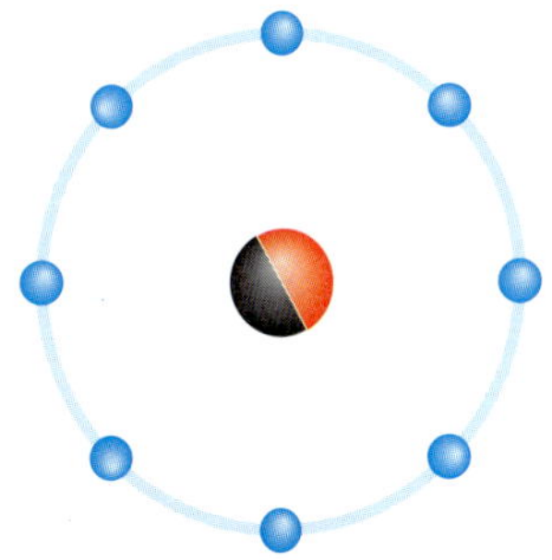

This complete shell means that the noble gases are very reluctant to take part in chemical reactions because that would tend to disturb this stable arrangement.

The noble gases are therefore chemically very inert and unreactive.

Uses of the noble gases

Helium:

- ☐ is less dense than air and does not burn. It can therefore be safely and effectively used in airships and weather balloons
- ☐ deep-sea divers breathe a mixture of helium and oxygen which does not contain the nitrogen which dissolves in the blood and can cause decompression sickness (the 'bends').

Neon:

- ☐ is used in advertising and warning signs because it glows brightly when electricity is passed through it.

Argon:

- ☐ is used in domestic electric light bulbs. They are filled with argon because its inertness prevents the filament from burning away
- ☐ is used to produce an inert atmosphere in processes such as welding and casting metal. In an atmosphere of argon the metal does not oxidise.

Q1 Explain why the melting points and boiling points of the noble gases increase with increasing atomic number.

Q2 Explain why the noble gases are inert and unreactive.

Q3 There is much more oxygen than argon in air. Why, then, is argon obtained before oxygen in the fractional distillation of liquefied air?

Q4 Why is helium used in airships?

Q5 Which properties of argon make it good for filling domestic light bulbs?

6 The first transition series – scandium to zinc

Physical properties

Name	Scandium	Titanium	Vanadium	Chromium	Manganese	Iron	Cobalt	Nickel	Copper	Zinc
Symbol	Sc	Ti	V	Cr	Mn	Fe	Co	Ni	Cu	Zn
Atomic number	21	22	23	24	25	26	27	28	29	30
Electronic configuration	2.8.9.2	2.8.10.2	2.8.11.2	2.8.12.2	2.8.13.2	2.8.14.2	2.8.15.2	2.8.16.2	2.8.17.2	2.8.18.2
Density (g cm^{-3})	2.99	4.50	5.96	7.20	7.40	7.87	8.90	8.90	8.94	7.15
Melting point (°C)	1540	1670	1800	1850	1245	1530	1490	1450	1083	420
Boiling point (°C)	2830	3290	3380	2260	1900	2750	2870	2730	2630	910

Chromium, iron, copper, and zinc are common metals. Here are photographs of the other six metals in the series.

Scandium. ▲

$_{22}$Ti

2.8.10.2

Titanium. ▲

Vanadium. ▲

Manganese. ▲

Cobalt. ▲

Nickel. ▲

The transition elements:

- ☐ are those elements whose atoms have an incomplete one-from-outermost (penultimate) electron shell
- ☐ lose electrons to form positive ions in chemical reactions and are therefore all metals
- ☐ are good conductors of electricity because the spaces in the penultimate shell let the electrons move
- ☐ have similar sized atoms because the number of electron shells in use is the same across the series. Moving left to right there is a trend to slightly smaller atoms because the increasing number of protons tends to attract the electrons in closer
- ☐ have a high density which tends to get higher across the series because the smaller atoms can pack closer together
- ☐ have strong inter-atomic forces because of the unfilled penultimate shell. This results in high melting and boiling points
- ☐ have considerable similarities to one another across the series because of their similar electronic configurations. This means that the kind of trends we saw as we moved down the group in the alkali metals, the halogens and the noble gases are not seen as we move across the transition series.

◀ In contrast to the metals of Groups I and II, which all have white compounds which form colourless solutions, the transition metals have coloured compounds which form coloured solutions in water. The reason for this is complex but it again involves the incomplete penultimate shell.

The position of zinc

Zinc is a special case in that it is included in the transition series in the Periodic Table but it has a complete penultimate shell. Zinc therefore has:

- white compounds and colourless aqueous solutions
- weaker inter-atomic forces and therefore lower melting and boiling points than the other metals in the series
- a slightly larger atom and therefore a lower density than copper.

Uses of the transition metals

Titanium has great strength, corrosion resistance and lightness and therefore:

- is used in spacecraft, aircraft and submarines
- is used in replacement hip joints and spectacle frames.

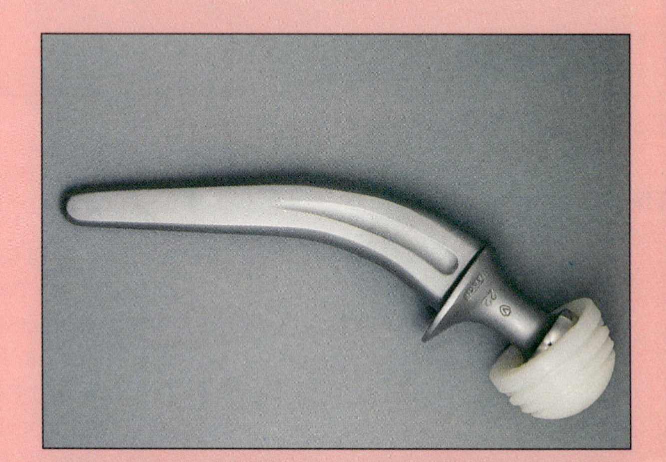

Chromium is extremely hard and resistant to heat and corrosion and therefore:

- is used in alloy steels for articles such as ball bearings and in plating steel for protection from corrosion.

Iron has a range of properties depending upon its purity and what it is mixed with. For example,

- wrought iron is a pure form of iron which can be softened by heat and then worked into shapes such as ornamental gates, chain links etc.
- cast iron is less pure than wrought iron. It is extremely hard but brittle and will break if struck hard. It is used for manhole covers, drainpipes, guttering, and machinery frames
- finely divided iron is used as a catalyst in the Haber process for the manufacture of ammonia (see Chapter 9).

Copper is both malleable (can be beaten into shapes) and ductile (can be stretched into wires), resistant to corrosion and an excellent conductor of both heat and electricity and therefore:

- is used for electrical wiring, both industrial and domestic
- is used for water pipes in domestic hot water systems, heat exchange units, the brewing industry and the food industry.

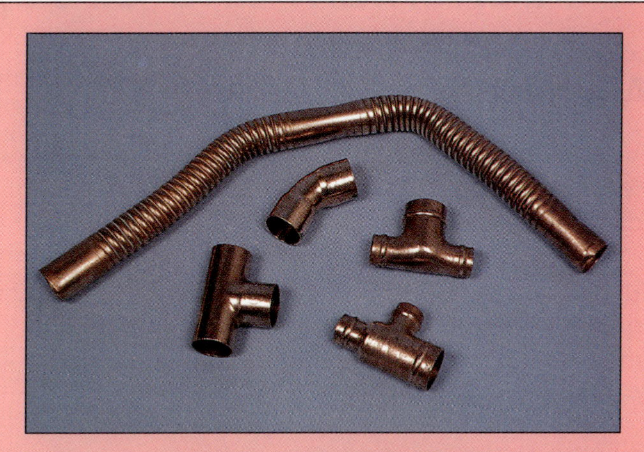

Zinc in its purest form is very resistant to atmospheric corosion and therefore:

- is used as a coating on iron or steel to prevent rusting. This coating of iron by zinc is called **galvanising**. The protection continues even after surface damage exposes the iron. This is because zinc is more reactive than iron and will therefore react with the atmosphere before the exposed iron does, thus maintaining the protection.

Q1 Explain why all the transition elements are metals.

Q2 Explain why the transition metals have very high melting and boiling points.

Q3 Explain why the density of the metals increases from left to right across the series up to copper.

Q4 Explain why the compounds of zinc are white while the other metals in the series have coloured compounds.

Q5 Tin is less reactive than iron. In that case, explain why the protection that tin coating gives to iron is not as good as that which galvanising provides.

Extension exercises 13 and 14 may be used now.

7 Aluminium and Group III

Aluminium

Aluminium is in group III of the Periodic Table which consists of the following elements: ▶

Name	Symbol	Atomic Number	Electronic Configuration
Boron	B	5	2.3
Aluminium	Al	13	2.8.3
Gallium	Ga	31	2.8.18.3
Indium	In	49	2.8.18.18.3
Thallium	Tl	81	2.8.18.32.18.3

Aluminium, the most abundant metal in the Earth's crust, is the most important of the group. It has a wide variety of uses.

The extraction of aluminium

The high reactivity of aluminium means that it can only be extracted from its compounds by electrolysis.

It is extracted from an aluminium ore called bauxite. Bauxite is aluminium oxide Al_2O_3 mixed with small amounts of iron oxide and sand.

The electrolytic cell for the extraction of aluminium from bauxite. ▼

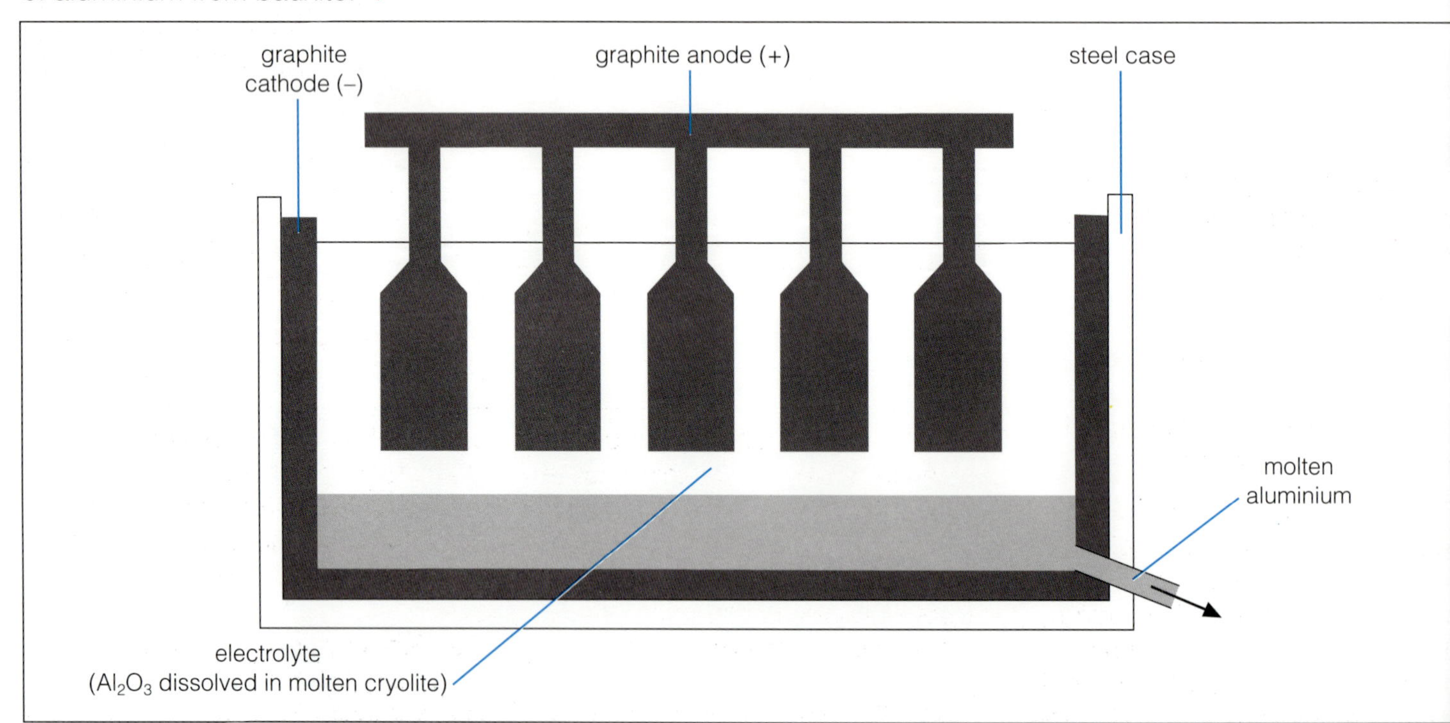

1. Removal of the impurities from the bauxite produces pure Al_2O_3 which has a melting point of above 2000°C. To electrolyse it at this temperature would be both dangerous and expensive.

2. The Al_2O_3 is dissolved in molten cryolite, a compound of sodium, aluminium and fluorine. It melts at about 900°C and the electrolysis is carried out at this temperature.

3. The reaction which occurs in the electrolysis is:

aluminium oxide → aluminium + oxygen

$$2Al_2O_3 \rightarrow 4Al(l) + 3O_2(g)$$

4. The aluminium is produced at the cathode and at 900°C is molten and so can be tapped off. The oxygen is produced at the anode where it is collected.

Reactivity of aluminium

Aluminium is a very reactive metal yet, surprisingly, if dilute hydrochloric acid is added to a small piece of aluminium foil there is no reaction at first. ▶

Wear eye protection.

The reaction begins very slowly before becoming the expected vigorous one. ▶

aluminium + hydrochloric acid → aluminium chloride + hydrogen

$$2Al(s) + 6HCl(aq) \rightarrow 2AlCl_3(aq) + 3H_2(g)$$

Your teacher may allow you to try this or may demonstrate it.

The reason for the initial lack of reaction is that the pieces of aluminium have become coated with a thin but strong film of aluminium oxide. Only when this film has dissolved in the acid can the acid get to the aluminium to produce the expected vigorous reaction.

This protective film is very useful. It can be strengthened by electrolysis. The aluminium is the anode and it becomes oxidised more. The process is called **anodising**.

Uses of aluminium

Aluminium is a low density, light metal; a good conductor of heat and electricity; malleable and ductile; non-toxic; very strong when alloyed with copper to make duralium (96% Al, 4% Cu); non-corrosive when the oxide film is left on or strengthened by anodising.

Some uses of aluminium which result from these properties are:

- ☐ in overhead electricity cables
- ☐ making saucepans
- ☐ making aircraft parts (duralium)

- ☐ making window frames, greenhouse frames, tent poles
- ☐ as thermal insulation in aluminium blankets which reflect heat back into the body
- ☐ in food packaging: milk bottle tops; baking foil; frozen food wrapping; drink cans.

Q1 Aluminium is extracted from bauxite. Explain why this is done by electrolysis.

Q2 Give the name and formula of the aluminium compound that is the main constituent of bauxite.

Q3 Explain why the purified bauxite has to be dissolved in cryolite before it is electrolysed.

Q4 Give a balanced chemical equation that summarises the industrial production of aluminium by electrolysis.

Q5 Explain why pieces of aluminium do not react immediately when dropped into dilute hydrochloric acid. Say what you would expect them to do.

Q6 State two properties of aluminium in each case which make it useful for:
a overhead electricity cables
b milk bottle tops
c tent poles
d aluminium blankets
e saucepans
f aircraft bodies.

Extension exercise 15 can be used now.

8 Carbon and Group IV

Structure, property and uses of diamond and graphite

Group IV of the Periodic Table consists of the following elements ▼

Name	Symbol	Atomic Number	Electronic Configuration
Carbon	C	6	2.4
Silicon	Si	14	2.8.4
Germanium	Ge	32	2.8.18.4
Tin	Sn	50	2.8.18.18.4
Lead	Pb	82	2.8.18.32.18.4

We shall look at carbon and, in particular, at diamond and graphite.

The allotropy of carbon

Carbon can exist in two different forms. One form is diamond and the other is graphite. Both are pure, solid carbon and both consist only of carbon atoms.

Diamond and graphite are called **allotropes** of carbon. *An element is allotropic when it can exist in different forms in the same physical state.* The different forms are the allotropes of the element.

The differences between diamond and graphite are caused by the different ways in which the carbon atoms are held together. This difference in structure gives them different properties.

Diamond

A carbon atom has an electronic configuration of 2.4. In diamond each carbon atom uses its four outermost electrons to bond to four others. This is repeated to make a large three-dimensional structure. It consists of millions and millions of atoms built up into one single giant molecule. This giant molecule is called a **macromolecule**.

Properties of diamond

- ☐ The bonds are very strong so diamonds are very hard. Diamond is the hardest natural substance known.
- ☐ The very strong bonds also give diamond a very high melting point of 3500°C.
- ☐ Because the structure is so rigid and all the outer electrons are used and firmly held, diamond does not conduct electricity.

Uses of diamond

- ☐ Because of its value and appearance diamond is used to make jewellery.
- ☐ Diamonds are used in glass cutters, the points of rock drills, medical instruments, record player styluses and, when powdered, as abrasives.

The central carbon atom is bonded to four others in a tetrahedral shape. ▲

The tetrahedral structure is repeated many times and builds into a macromolecule. ▲

The macromolecule is large enough to produce a diamond. ▲

Graphite

In graphite each carbon atom uses three of its outer electrons to bond to three other carbon atoms in a triangular pattern. This produces a structure that is a flat sheet in which each carbon atom is bonded to three others and has a spare electron.

Each atom is bonded to three others in a flat sheet. ▲

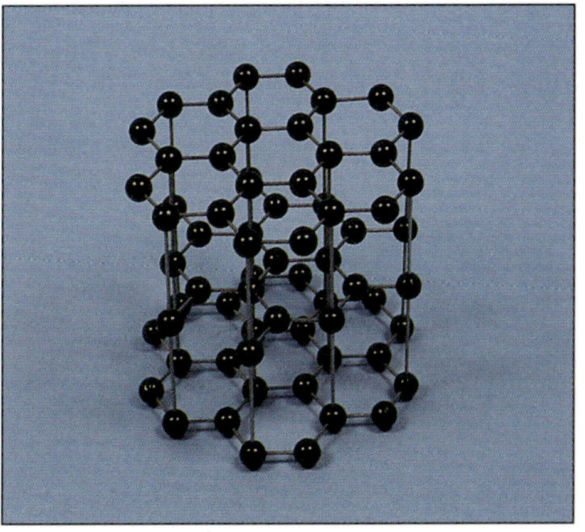

The sheets lie one on top of another. ▲

The layered macromolecules produce graphite where the layers can be seen. ▲

A piece of graphite consists of millions and millions of these sheets lying one on top of another with very weak forces between the sheets.

Properties of graphite

- ☐ The forces between layers are weak and the layers are about two to three times further apart than the bonded atoms within the layer so the layers slide over each other easily. Thus graphite is dark grey, has a dull shine and feels greasy and slippery.
- ☐ The bonded atoms within the layer are fixed in place very strongly so the melting point of graphite is extremely high at 3700°C.
- ☐ The spare electron on each carbon atom is free to move along the layer from one atom to the next and therefore graphite is a very good conductor of electricity. It is the only non-metal that conducts electricity.

Uses of graphite

- ☐ Because the layers slide over one another so easily, graphite is used as a lubricant and as the 'lead' in pencils.
- ☐ Its high heat resistance makes it suitable for making crucibles used to hold molten metal.
- ☐ Its electrical conductivity makes it suitable for use as an electrode in electrolytic processes, as the anode (+) in domestic batteries, and in resistors.
- ☐ Because the bonds within layers are so strong but those between layers are much weaker, graphite fibres are both strong and flexible as the layers slide. Carbon fibre is used in golf clubs, tennis racquets, vaulting poles, etc. where strength and flexibility are necessary.

Q1 Explain what is meant by an 'allotropic element'.

Q2 Explain the structural differences between the two allotropes of carbon.

Q3 Why do both diamond and graphite have extremely high melting points?

Q4 Explain why graphite will conduct electricity but diamond is a non-conductor.

Q5 Which property of graphite makes it suitable for pencil 'lead'?

Q6 What is a macromolecule?

Q7 Why is carbon fibre used in the shafts of golf clubs?

Extension exercise 16 can be used now.

9 Nitrogen and Group V

Nitrogen: properties and the Haber process

Group V of the Periodic Table consists of the following elements:

Name	Symbol	Atomic Number	Electronic Configuration
Nitrogen	N	7	2.5
Phosphorus	P	15	2.8.5
Arsenic	As	33	2.8.18.5
Antimony	Sb	51	2.8.18.18.5
Bismuth	Bi	83	2.8.18.32.18.5

We shall look only at nitrogen which is a colourless, odourless, tasteless gas making up 78 per cent by volume of the atmosphere.

- ☐ The boiling point of nitrogen is –196°C and it is obtained from air by fractional distillation (see Chapter 5).
- ☐ Nitrogen exists as diatomic molecules (N_2). Because the bonding between the two atoms is very strong, nitrogen is relatively inert and unreactive.
- ☐ The most important use of atmospheric nitrogen is its combination with hydrogen to produce ammonia:

nitrogen + hydrogen → ammonia

$$N_2(g) + 3H_2(g) \rightarrow 2NH_3(g)$$

The Haber process

The Haber process is the most important method of manufacturing ammonia. Fritz Haber first discovered the process in 1908.

The raw materials are natural gas (methane, CH_4), air and water. The stages of the process are:

1 Natural gas and steam are passed into a converter at a pressure of 30 atmospheres and a temperature of 900°C and with a catalyst of nickel. A mixture of carbon dioxide and hydrogen is formed:

natural gas + steam → carbon dioxide + hydrogen

$$CH_4(g) + 2H_2O(g) \rightarrow CO_2(g) + 4H_2(g)$$

2 Air is added. The oxygen in the air reacts with some of the hydrogen to form steam:

$$2H_2(g) + O_2(g) \rightarrow 2H_2O(g)$$

3 The mixture now contains carbon dioxide, steam, nitrogen and hydrogen. Passing this through a concentrated solution of potassium carbonate removes the carbon dioxide and the steam.

Finely divided iron is used as the catalyst in the Haber process. ▶

The ammonia manufacturing plant in Billingham. ◀

4 This leaves nitrogen and hydrogen, the two required reactants. These are reacted together at a pressure of at least 250 atmospheres in the presence of a finely divided iron catalyst at a temperature of 500–550°C:

$$N_2(g) + 3H_2(g) \xrightarrow[\text{500–550°C, Fe catalyst}]{\text{250 atm}} 2NH_3(g)$$

5 About 15 per cent of the gases are converted to ammonia which is removed as a liquid by cooling under pressure. Unreacted nitrogen and hydrogen are recycled to make more ammonia.

Fertilisers

Nitrogen is an essential constituent of all living matter. Nitrogen compounds, proteins in particular, are needed for the healthy development of all plants and animals.

Animals get their protein from the food they eat. Plants get protein by taking nitrogen compounds from the soil and converting them to protein.

However, if the same soil is continuously used for planting and re-planting then its nutrients will eventually be used up and they will need to be replaced by adding fertilisers to the soil.

Milk is a high protein food consumed by all young animals. ▲

Fertilisers produce high yields of healthy crops over long periods of time. ▲

Ammonium salts, such as ammonium nitrate and ammonium sulphate, are excellent **nitrogenous** fertilisers. They are manufactured by reacting ammonia with dilute sulphuric or dilute nitric acid:

ammonia + nitric acid → ammonium nitrate

$$NH_3(g) + HNO_3(aq) \rightarrow NH_4NO_3(aq)$$

ammonia + sulphuric acid → ammonium sulphate

$$2NH_3(g) + H_2SO_4(aq) \rightarrow (NH_4)_2SO_4(aq)$$

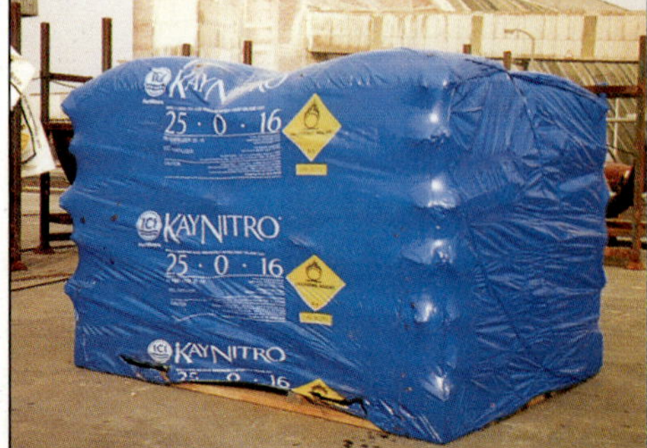

Fertiliser production in the UK can reach 10 000 tonnes per day. ▶

Q1 Name the raw materials used in the Haber process for the manufacture of ammonia.

Q2 Explain, with equations and conditions, the chemistry of the conversion of these raw materials to the nitrogen and hydrogen needed for the Haber process.

Q3 Explain, with a balanced chemical equation, how nitrogen and hydrogen are converted into ammonia in the Haber process.

Q4 Reacting nitrogen with hydrogen is difficult because the nitrogen is 'chemically inert'. Explain what this means and why it applies to nitrogen.

Q5 How do animals get protein for healthy growth?

Q6 How do plants get protein for healthy growth?

Q7 Explain what is meant by 'nitrogenous fertiliser'. Name two examples.

Q8 Write balanced chemical equations for the conversion of ammonia gas into two nitrogenous fertilisers.

Extension exercise 17 can be used now.